ISAAC ASIMOV'S
Library of the Universe

Colonizing

the

Planets

and

Stars

by Isaac Asimov

Gareth Stevens Publishing
Milwaukee

Library of Congress Cataloging-in-Publication Data

Asimov, Isaac, 1920-
  Colonizing the planets and stars / by Isaac Asimov. — A Gareth Stevens
Children's Books ed.
        p.   cm. — (Isaac Asimov's library of the universe)
  Bibliography: p.
  Includes index.
  Summary: Explores the possibility of establishing colonies in space, traveling
by starship to other galaxies, and meeting extraterrestrials.
  ISBN 1-55532-372-3
  1. Space colonies—Juvenile literature. [1. Space colonies.] I. Title. II. Series:
Asimov, Isaac, 1920-   Library of the universe.
TL795.7.A85   1989
919.9'04—dc20   89-4644

A Gareth Stevens Children's Books edition

Edited, designed, and produced by
Gareth Stevens, Inc.
RiverCenter Building, Suite 201
1555 North RiverCenter Drive
Milwaukee, Wisconsin 53212, USA

Cover painting © Rick Sternbach

Project editor: Mark Sachner
Series design: Laurie Shock
Book design: Kate Kriege
Research editor: Kathleen Weisfeld Barrilleaux
Picture research: Matthew Groshek
Technical advisers and consulting editors: Julian Baum and Francis Reddy

Printed in the United States of America

2 3 4 5 6 7 8 9 95 94 93 92 91 90

CONTENTS

Nowadays, we have seen planets up close, all the way to distant Uranus and Neptune. We have mapped Venus through its clouds. We have seen dead volcanoes on Mars and live ones on Io, one of Jupiter's satellites. We have detected strange objects no one knew anything about till recently: quasars, pulsars, black holes. We have studied stars not only by the light they give out but by other kinds of radiation: infrared, ultraviolet, x-rays, radio waves. We have even detected tiny particles called neutrinos that are given off by the stars.

But now, it does not seem enough to look at things. Human beings have left our planet and returned, and it may be that in the next century, people will be living here and there out in space. Where will they go? How will they manage out in space? In fact, why should they go? Here, in this book, we will take up the subject of settling the cosmos.

Isaac Asimov

Why Should We Go?

Most scientists agree that the first human ancestors appeared in Africa about two million years ago. Our ancestors extended their range, moving outward, and now human beings — five billion of us — live just about everywhere on this planet. The urge to move outward is still with us after all this time — and today it extends beyond Earth, to space.

Why do we dream of reaching out to the cosmos? There are many reasons. Many people worry that one day we will use up Earth's natural resources. Many feel that chemicals we release into the atmosphere will destroy Earth's protective ozone layer. Others simply feel that humans will always have the need to settle new regions, even if it means going into space.

No other place suits us in the way Earth does, but perhaps we can make other worlds livable. Perhaps, one way or another, we can thrive out there beyond Earth.

Below: The oldest human fossils appear in central Africa. From there our ancestors slowly migrated to new lands all across the globe. Perhaps one day our migration will take us to other worlds.

Opposite: Our earliest ancestors gazed at the Moon and stars with fear and wonder.

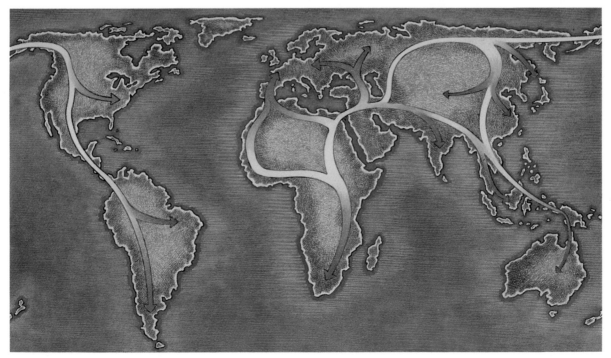

Background: a slice of life in an ant colony.

Above: Ant colonies contain thousands of insects all working for the common good of the group. Cooperation will be as important to human space colonies as it is to earthly ants.

The test module for Biosphere II, the most ambitious plan to date for creating a completely closed ecology. This work will help space planners create environments for space stations and colonies on the Moon and Mars.

Making a Start

How do we begin a venture as mammoth as settling the vast regions of space?

It would take a great deal of cooperation and planning, but to begin with, we might build worlds of our own in space. We might build cylinders or spheres, and inside these we can construct buildings, roads, lakes, and so on. Such artificial settlements could easily hold 10,000 people each. One scientist has even worked out designs for settlements that would each hold 10 million people — big enough for all of New York City!

Such settlements would rotate, so people would be pushed against the walls as if there were gravity. There would be farms and factories, and no bad weather.

The Sun would light the settlements and supply energy, and we could create the impression of day and night through various clever tricks. One would be to open and close huge shades. Another would be to place rotating mirrors outside the settlements. These mirrors would reflect the entire image of the Sun onto the settlement to create "daylight."

Below, left: Skylab, NASA's first space station, helped us learn more about the way our bodies respond to long periods in space. Right: Soviet cosmonauts have stayed in space for nearly a year.

The First World Bases — The Moon and Mars

As we set our sights on other worlds, we will need the Moon as a mining base. The Moon's soil can be used to make metals, glass, oxygen, and other substances we would need to build settlements and structures in space. Later, we can build cities underground on the Moon, filling them with air and bringing in water from Earth.

Mars will come later. It is bigger than the Moon and has its own water. It, too, will have underground cities. People in settlements on the Moon and Mars will become used to living inside structures. They will be prepared for longer trips than Earth people could manage.

Above: Water is one ingredient we'll need
anywhere we go. Water vapor condenses
into drops and rains from the sky, to be
used by plants and animals.

Opposite: an artist's idea of an outpost on Mars.

Below: Four US scientists stayed in this underwater habitat for three months. They
showed that people could live and work in an alien environment.

Life among the Asteroids

Beyond Mars lies the asteroid belt, with perhaps as many as 100,000 asteroids orbiting the Sun between Mars and Jupiter. The asteroids might be a wonderful resource for cosmic colonizers of the future. For one thing, the asteroids could be the largest source of minerals anywhere in the Solar system. In addition, even the small ones could be hollowed out to make worlds larger than any we can build in space for ourselves.

Someday there may be many thousands of settlements in the asteroid belt, each with a million people or more, each with its own customs and culture. There might be as many people in the asteroids as on Earth, and it may be the "asteroid people" who will make the long trips to the outer Solar system.

Opposite: Future colonists create a spaceship from a hollowed-out asteroid. Inset: Astronauts ride an asteroid back to Earth, where it will be mined for useful materials.

Below: Our bodies have adapted to the daily cycle of Earth's rotation. Scientists want to study the "biological clocks" within us. In one famous experiment, Stefania Follini spent over 130 days alone deep within a cave. When she emerged, her body had formed new rhythms. What will happen to humans who spend years in space?

Variety makes the world (and cosmos) go 'round

But is variety something people will want when they start a space colony and leave Earth forever? Who will go? Who will stay behind? People with certain talents and skills will have to go. Would you want people who would be homesick for the Solar system, or frightened by the vast emptiness between the stars? Must everyone be exactly alike? How would you choose the right people and build a healthy starship society? Right now, nobody really knows.

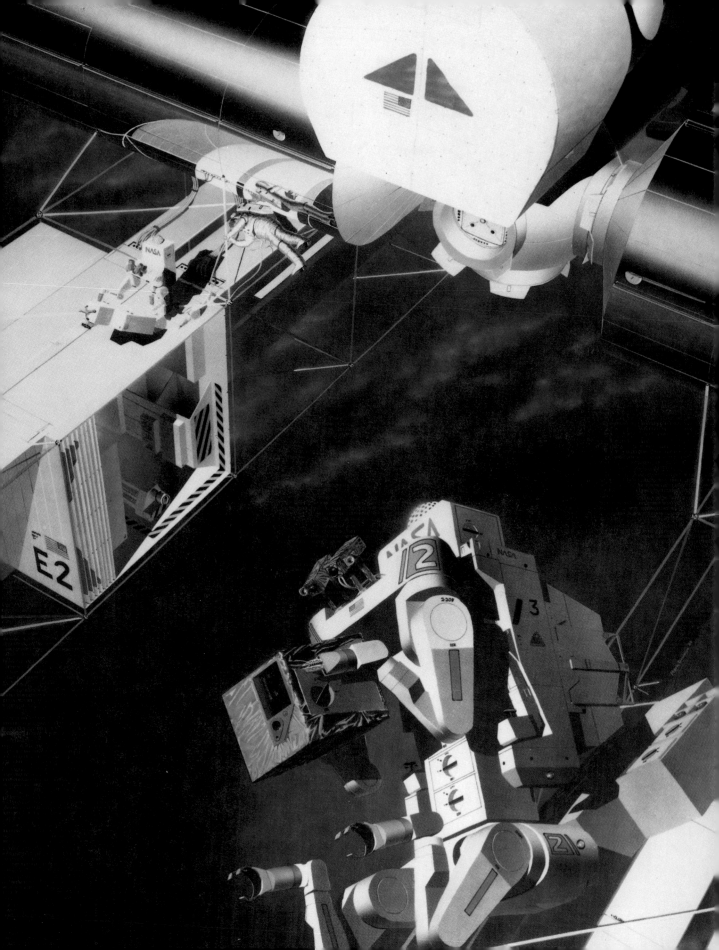

Life in the
Solar Neighborhood

With people living in space, space travel will become quite common. Many spaceships will be built in space and will coast easily from one place to another because they won't have to fight Earth's gravity to do so. There will be ships carrying goods from one settlement to another, rescue ships in case of wrecks, and repair ships to keep space settlements and underground cities in good shape.

Many of these ships could be run by robots equipped to do simple jobs. Robots could also run factories in space and could work at solar power stations that extract energy from the Sun.

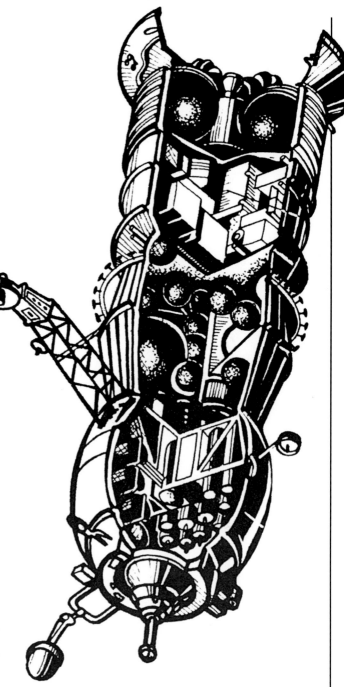

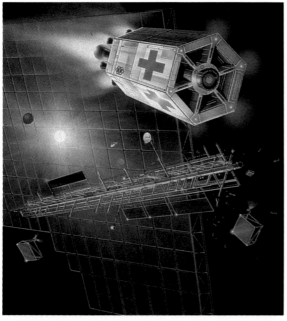

Above: Soviet Progress ships brought fresh supplies to orbiting cosmonauts. The unpiloted vehicles were controlled from the ground.

Left: a robot space ambulance.

Opposite: Robot helpers assist astronauts in the construction of a space habitat.

Getting Around on Other Worlds

The Moon's surface is almost as large as North and South America combined. Mars's surface is almost as large as the total land area of Earth. Once these worlds are settled, subways may carry people from one city to another. They would run in a vacuum, and the cars would speed along on magnetized rails at hundreds of miles an hour.

Walking on other worlds might be easier, too, where gravity is lower. Mars has only about two-fifths Earth-gravity. The Moon has only one-sixth Earth-gravity. With lower gravity, muscles might weaken, so people would work out to stay in shape. But at low gravity, exercise would also be different, and even gymnastics might be fun.

Opposite: Future colonists on Mars may develop a "Mars-way" — an underground subway system between outposts. Inset: Dirigibles could give Mars colonists a unique view of their adopted planet.

This "Moonbuggy," called the Lunar Rover, was taken to the Moon on each of the last three US Apollo missions. It gave astronauts much more mobility on the lunar surface.

A fleet of space cities prepares to explore the outer reaches of the Solar system.

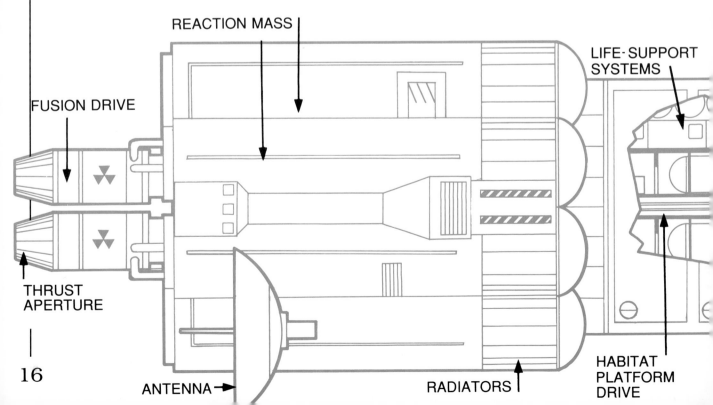

REACTION MASS

LIFE-SUPPORT SYSTEMS

FUSION DRIVE

THRUST APERTURE

ANTENNA

RADIATORS

HABITAT PLATFORM DRIVE

To the Edge of the Solar System . . .

The outer planets lie far beyond the asteroid belt, so it might take years to reach them. But the asteroid people could make the trip livable by building large ships with crews of hundreds. Such ships would be little worlds, much like the asteroids, so the crew would feel at home and wouldn't mind the trip.

We might be able to settle some of the moons of the outer planets, though the gas giants themselves — Jupiter, Saturn, Uranus, and Neptune — are far too alien for human use. Eventually, we may even build an outpost on distant Pluto, where the Sun would only seem a brightly glowing star. There, from the rim of our Solar system, people will look outward toward the other stars.

Below: a blueprint for a piloted ship to be used in the long-range exploration of the Solar system. Perhaps such a vessel will one day take astronauts to distant Pluto.

Inset: Astronauts prepare to visit Charon, the moon of Pluto.

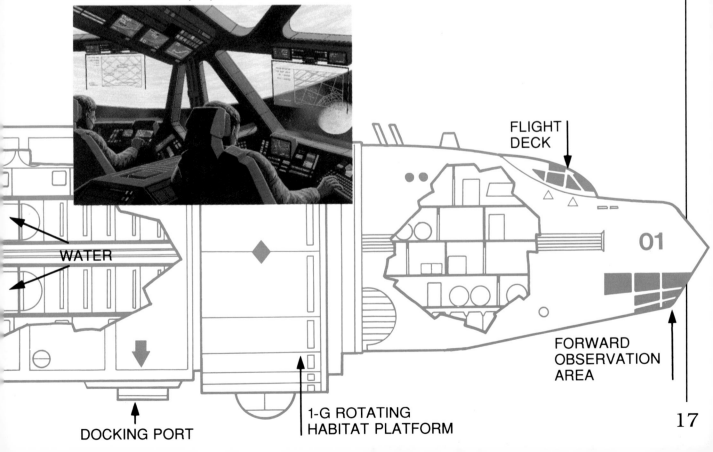

WATER

FLIGHT DECK

01

DOCKING PORT

1-G ROTATING HABITAT PLATFORM

FORWARD OBSERVATION AREA

. . . And Beyond

What happens when the Solar
system is completely settled by
humans? Where do we go next?
There are many trillions of other
stars, many of which must have
planets orbiting them. But can
we reach them? Even the nearest
star is about 7,000 times as far
away from us as Pluto is.

The speed of light is about
186,000 miles (300,000 km) per
second. That seems very speedy
— faster than anything we can
imagine here on Earth. But even
at the speed of light, it would take
over four years to reach Alpha
Centauri, the nearest star. It
would take even longer to reach
farther stars, and, at the speed of
light, 100,000 years to go from
one end of our Milky Way Galaxy
to the other! It would take well
over 100,000 years to reach our
closest galactic neighbors, and
millions of years to go on to
farther galaxies — even at the
speed of light.

A starship blasts out of Earth orbit en route
to the stars.

The distance between stars in our Milky Way Galaxy will keep us in the neighborhood of our Sun for a long time. Perhaps one day, though, we'll want to take another great leap — to a nearby galaxy. The nearest galaxy like our own is so far away that its light takes over two million years to reach us.

**Other galaxies —
so "near" and yet so far**

Imagine that we've filled every star system in our Galaxy that might have planets fit for humans. Must we stop then? Not necessarily. There are other galaxies. For instance, three small galaxies, the Magellanic Clouds, are over 150,000 light-years away. The mammoth Andromeda Galaxy is 2.3 million light-years away. It would take millions of years to reach them, but perhaps we can do it if our descendants build starships capable of drifting long enough.

Star Travel — What Will It Take?

How do we build up enough energy to make trips even to the nearby stars? Instead of using rocket fuel, we could use ion drives to push tiny charged atoms (or "ions") backward. That would slowly but surely add speed more efficiently than ordinary engines would. Nuclear power could also be used to drive our ship.

We could even push our ship with laser beams. But no matter how much energy we use, it will still take many years to reach even the nearby stars.

The Bussard Ramjet Starship: an artist's concept of a craft that "scoops" hydrogen out of space to be used as fuel.

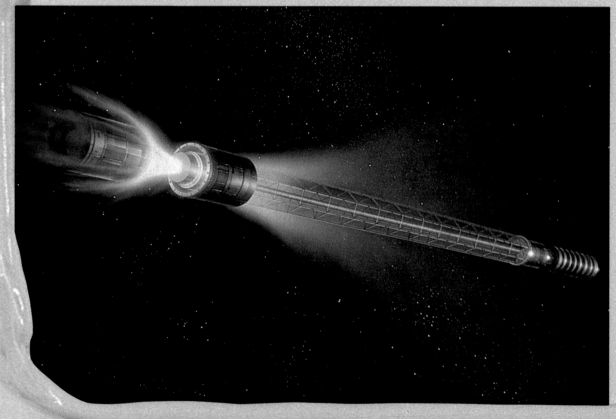

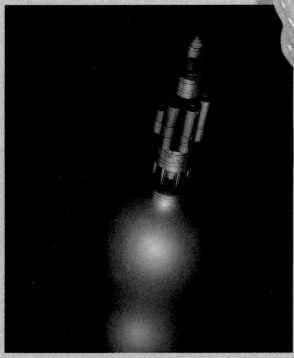

The Orion Nuclear-pulse Craft: Had this 1950s concept become reality, nuclear explosions would have propelled craft through space.

A galaxy of wonders — step right up!

Many cosmic oddities await curious star travelers. Imagine encountering a white dwarf, an incredibly dense object as massive as a star but smaller than Earth. A neutron star is also as massive as a star but less than 10 miles (16 km) across. Its gravitational pull is so terrific you wouldn't want to come too close. Imagine encountering a supergiant, a star that is hundreds of millions of miles wide. The Universe is full of strange and marvelous sights.

Imagine how much toothpaste it would take to keep your teeth clean between Earth and the stars. It would leave a trail of astronomical proportions!

SUPER NOVA-BRITE

Generations in Space

Of course, space settlers might not care about moving quickly through the cosmos. They might want to take their time. A whole asteroid settlement might decide to become a "starship" and leave the Solar system. They would outfit their settlement with advanced ion drives and take off, coasting at speeds of mere hundreds of miles a second. They would be in their own world, living their own lives, while it took their settlement many thousands of years to reach other worlds.

Many generations would be born and would die before they reached another world, but that wouldn't matter. They wouldn't have left home. Home would have come along with them.

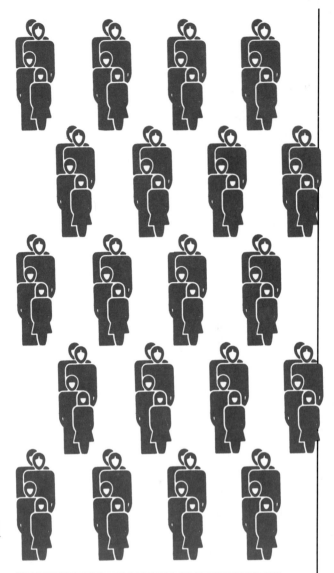

A galactic baby boom?

A starship in flight can only support so many people, so it must control its population. Once another star is sighted with a planet like Earth, or with an asteroid belt, things would be different. People would settle the planet or build new asteroid homes. They could then increase their population. When, after thousands of years, the new worlds are full, another group of starships might set out. Eventually, the entire Galaxy might be settled this way.

Opposite: Mobile living units on an asteroid would make city building easier. Inset: A "mobile living unit" on Earth.

Cosmic Jet Lag — The Star-Crew System

Let's imagine that such starships became common and slowly spread out from the Solar system in every direction. These ships might want to stay in touch with each other, and with Earth. To do this, the starship residents might use special ships that travel at nearly the speed of light.

The crews on those ships would find that time would travel slowly for them. A star crew might reach a spaceship, deliver a message, and return to Earth feeling as though only a few weeks had passed, while on Earth a century may have passed.

Earth travelers have "jet lag" when they change time zones. Think of the incredible adjustments to a new time that star crews would have to make. And think, too, of the shock star crews would have at losing everything they have known, such as family and friends. Working on a star crew could be one of the loneliest, most difficult jobs in the Universe!

Opposite: During the long trips to other stars, the crews of starships would have to fight extreme boredom. Here crew members entertain themselves with a card game.

Below: A starship approaches one of the new worlds in its new solar system.

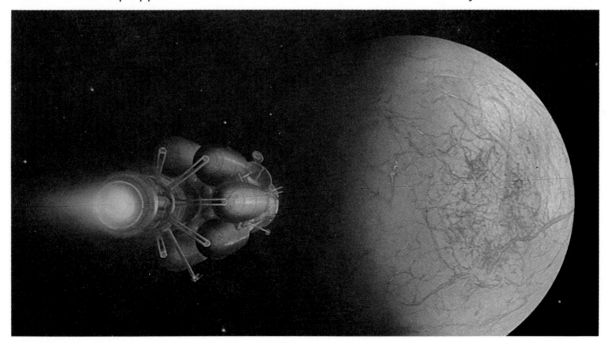

As the continents of our planet slowly separated, isolated land areas like Australia developed unique forms of life. Kangaroos (left), koalas (below), and the duck-billed platypus (above) exist in the wild nowhere else in the world. As humans expand into the Galaxy, each colony will develop differently. When descendants of different colonies meet millions of years in the future, they may not realize they're related.

Return to Earth?

As humanity spreads out through the Galaxy over millions of years, the star-crew system simply might not work, and people in one star system might know nothing about people in others. They may all evolve in different directions, populating the Galaxy with millions of different kinds of people.

Ten million years from now, while exploring a certain solar "neighborhood," strange people of the future might come across Earth. It, too, will have changed tremendously. Will these newcomers from space ever know they were home — that they were back on the planet where it all started? Perhaps not.

Earth's surface as it appeared 200 million years ago (top) and today (bottom). What were once connected "rafts" of land are today distinct continents.

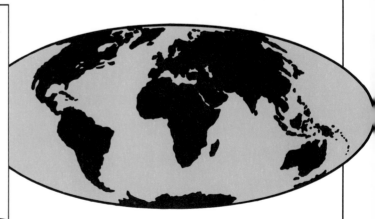

Are we alone?

Can we assume we are the only beings in the Galaxy with our kind of intelligence? We don't know, but it's hard to believe there aren't others elsewhere among all the hundreds of billions of stars in the Galaxy. Will our starships reach worlds with intelligent beings on them? What if we encounter starships with nonhumans? We might find them so different that we can't deal with them. Or we might find we can learn a great deal from them. Which will it be?

27

Fact File: Settling the Cosmos

Is your curiosity cosmic? If so, the idea of settling planets and distant star systems might appeal to you. The chart on these pages gives you an idea of just a few of the things you'd have to consider before joining up to settle space.

Before attempting to settle the Moon, Mars, and interstellar space, Earthlings may want to set up Earth-orbiting colonies even closer to home. Like space stations, these colonies would give people a place to work and play just beyond Earth's atmosphere. The purpose of such colonies might be to set up research labs, relay stations, and even factories in space. Beyond the Moon and Mars, and before taking the leap into deep space, we might want to settle the

Colonizing the Moon, Mars, and Beyond — Some Tricks of the Trade

Where To?	Travel Time?	Gravity?
Earth's Moon	One-way trip from Earth: several days. Communications with Earth: 3 seconds for round-trip transmission.	One-sixth that of Earth's, so daily exercise routine necessary. Will be difficult to rotate colony to provide artificial gravity.
Mars	One-way trip from Earth: 6-9 months. Communications with Earth: 10-40 minutes for round-trip transmission.	Two-fifths that of Earth's. Like Moon colony, Mars colony will be based on ground, difficult to rotate. Exercise regimen a must.
The Stars	One-way trip from Earth: indefinite. Closest star, Alpha Centauri, is "only" 4.3 light-years away, but will we ever be able to travel that fast? Trip might have to be made in stages, over many generations. Communications with Earth: years; eventually, none likely.	Artificial gravity possible by rotating deep-space colony. Also possible to adjust to gravity of final destination. Travelers will have years, possibly generations, to adapt to "new" gravity.

LUNA-GLOW

MARVO

STARBRITE

asteroid belt or some other part of the outer Solar system. As with the Moon or Mars, mining might be a good reason to set up shop among the asteroids. Or we might be out there to relieve overcrowding or the taxing of natural resources on Earth.

All this sounds pretty exciting. But it would also be a huge challenge. Think of having to create an artificial atmosphere, or of harnessing Solar, nuclear, and other forms of energy, of producing "natural" resources in space, of controlling the environment in general over an incredibly long time. The demands of time and energy might seem too great! But if we move step by step, starting with Earth-orbiting colonies and lunar colonies, both time and science could be on our side and propel us to the stars.

Help?	Nonhuman Inhabitants?	Civilization in Space?
Medical: Ferrying of medicine from Earth should be fairly routine. Colony will need its own doctors, nurses, and other medical staff. Other: Within a few days' travel time, supplies and other help are easily available from Earth.	Plants, animals, insects, and useful bacteria for farming and recycling air in permanent colony. Closeness to Earth would make experimenting and bringing in new varieties easy.	Because of proximity to Earth and ease of communication and travel between Earth and Moon, early settlers would probably be tied to Earth cultures from which they came. In time, as on Earth, settlements might increasingly take on their own character, along with their own distinct cultural identity.
Medical: Because of greater distance from Earth, any medical help that can't wait 6-9 months should be in-colony. Dentists now more crucial as low gravity can cause bone deterioration. Other: Long waits mark Earth-to-Mars shipping. Nearby space colonies likely drop-off points.	As on Moon, plants, animals, insects, and helpful bacteria for farming, breeding, and recycling. One big difference: Life forms would be harder to replace because of distance from Earth.	As in Moon settlements, Mars colonists would be tied to Earth by fairly quick communications. But travel time and proximity to the asteroids and beyond might isolate settlers. Over time, it's possible that "Martian" communities would develop their own identity as a society and even form their own governments.
Medical: Unlikely that deep-space settlements will have contact with — or even knowledge of — each other. Thus, all medical help must come from within each colony. Other: All help must come from within colony. Even "quickest" communication would take years.	There's no going back! Thus any mistakes in combinations of plants, animals, insects, or bacteria could be deadly to the community. Must be sure of the right combination from the start.	Schools, medical books, farming tools, music, computers, factories, research labs, libraries, leisure activities and equipment, art, hospitals: All this and more — or the know-how to develop it — must be brought along. But after generations in space, the colony would evolve into a civilization unlike any known on Earth.

More Books About Colonizing Space

Here are more books about colonizing space. If you are interested in them, check your library or bookstore.

Asimov on Astronomy. Asimov (Doubleday)
Colonies in Orbit. Knight (Morrow)
The High Frontier. O'Neill (Morrow)
How Do You Go to the Bathroom in Space? Pogue (TOR Books)
Out of the Cradle: Exploring the Frontiers Beyond Earth. Hartmann (Workman)

Places to Visit

You can explore the Universe — including the places where colonies may be established beyond Earth — without leaving our planet. Here are some museums and centers where you can find many different kinds of space exhibits.

Burke-Gaffney Planetarium
Saint Mary's University
Halifax, Nova Scotia

Newark Museum Planetarium
Newark, New Jersey

University of British Columbia Observatory
Vancouver, British Columbia

Kirkpatrick Planetarium
Oklahoma City, Oklahoma

State Museum of Pennsylvania Planetarium
Harrisburg, Pennsylvania

Cincinnati Museum of Natural History
    and Planetarium
Cincinnati, Ohio

For More Information About Space Travel

Here are some places you can write to for more information about venturing into space. Be sure to tell them exactly what you want to know about. And include your full name and address so they can write back.

For information about planetary missions:
NASA Jet Propulsion Laboratory
Public Affairs 180-201
4800 Oak Grove Drive
Pasadena, California 91109

For a price list and information about space food (send a self-addressed, stamped envelope):
American Outdoor Products
1540 Charles Drive
Redding, California 96003

For catalogs of slides, posters, and other astronomy material:
AstroMedia Order Department
21027 Crossroads Circle
Waukesha, Wisconsin 53187

Sky Publishing Corporation
49 Bay State Road
Cambridge, Massachusetts 02238-1290

Selectory Sales
Astronomical Society of the Pacific
1290 24th Avenue
San Francisco, California 94122

Hansen Planetarium
15 South State Street
Salt Lake City, Utah 84111

Glossary

alien: in this book, a being from some place other than Earth.

asteroid: "star-like." The asteroids are very small planets made of rock or metal. There are thousands of them in our Solar system, and they mainly orbit the Sun in large numbers between Mars and Jupiter. But some show up elsewhere in the Solar system — some as meteoroids and some possibly as "captured" moons of planets such as Mars.

asteroid belt: the space between Mars and Jupiter that contains thousands of asteroids.

atmosphere: the gases that surround a planet, star, or moon.

bacteria: the smallest and simplest forms of cell life. A bacterium is one-celled and can live in soil, water, air, food, plants, and animals, including humans.

billion: in North America — and in this book — the number represented by 1 followed by nine zeroes — 1,000,000,000. In some places, such as the United Kingdom (Britain), this number is called "a thousand million." In these places, one billion would then be represented by 1 followed by *12* zeroes — 1,000,000,000,000: a million million, a number known as a trillion in North America.

evolve: to develop or change over a long period of time.

fertilize: to make something able to produce — in soil, the addition of manure or chemicals to make it able to grow plants.

generation: the average period of time between the birth of parents and the birth of their children.

gravity: the force that causes objects like the Earth and Moon to be attracted to one another.

ion drive: an engine that works by the ejection of charged atoms or molecules.

laser: L(IGHT) A(MPLIFICATION) by S(TIMULATED) E(MISSION) of R(ADIATION). A device that focuses light to a beam intense enough to burn holes through the hardest metals known.

Magellanic Clouds: the galaxies nearest the Milky Way, irregular in form and visible to the naked eye in the Southern Hemisphere.

Milky Way: the name of our Galaxy.

natural resources: materials supplied by the environment.

neutron star: a star with all the mass of an ordinary large star but with its mass squeezed into a much smaller ball.

oxygen: the gas in Earth's atmosphere that makes human and animal life possible. Simple life forms changed carbon dioxide to oxygen as life evolved on Earth.

ozone layer: that part of our atmosphere that shields us from the Sun's dangerous ultraviolet rays.

Pluto: the farthest known planet in our Solar system and one so small that some believe it to be a large asteroid.

red giant stars: huge stars that may be over a hundred million miles (160 million km) in diameter.

rotate: to turn or spin on an axis.

Solar system: the Sun with the planets and all other bodies, such as the asteroids, that orbit the Sun.

subway: an underground passage.

vacuum: a space without any matter, lacking even air.

white dwarf: the small white-hot body that remains when a star like our Sun collapses.

Index

The publishers wish to thank the following for permission to reproduce copyright material: front cover, pp. 13 (left), 20 (lower), 21 (upper), © Rick Sternbach; pp. 4, 5, © Keith Ward, 1989; p. 6 (lower), © 1989 Space Biospheres Venture; p. 6 (upper right), Runk/Schoenberger from Grant Heilman Photography; pp. 7 (left), 15, 27 (both), courtesy of NASA; pp. 7 (right), 13 (right), James Oberg Archives; p. 8, courtesy of NASA, artwork by Pat Rawlings; pp. 9 (upper), 20-21 (background), 28-29 (all), © Gareth Stevens, Inc., 1989; p. 9 (lower), official US Navy photograph; p. 10 (large), © David Hardy; p. 10 (inset), © Mark Maxwell, 1986; p. 11, © Mark Williams, Current Argus Photo, 1989; p. 12, courtesy of McDonnell Douglas Astronautics Company; pp. 14 (large), 16-17 (lower), © Mike Stovall, 1989; p. 14 (inset), © Michael Carroll, 1985; p. 16 (upper), © Julian Baum; p. 17 (inset), © Mark Dowman, 1989; p. 18, © Michael Carroll; p. 19, © Julian Baum, 1989; p. 22 (large), © Doug McLeod, 1989; pp. 22 (inset), 23 (upper left), courtesy of Airstream, Inc.; p. 24, © Mario Macari, 1989; p. 25, © Mark Maxwell, 1983; p. 26 (all), © John Cancalosi/Tom Stack and Associates.